Medical Evidence for

Hypnosis

By
Esmilda Abreu-Hornbostel
And
Matthew Craver

Table of Contents

Overview

Despite a long history of success, hypnosis still has limited acceptance in the medical community. This is unfortunate, because hypnosis can assist in the management and treatment of a number of conditions.

Hypnosis struggles for acceptance partly because of the image many people have of hypnosis; an image formed mostly by movies, TV, and stage hypnotists. Clinical hypnosis has little, if anything, in common with the hypnosis presented in these settings. In contrast, clinical hypnosis is supported by a large body of research in peer-reviewed scientific journals. In fact, the National Institute of Health's MEDLINE database indexes almost 500 clinical trials on hypnotherapy.

Clinical hypnosis is best thought of as a complementary therapy technique. It doesn't claim to cure conditions or substitute for traditional medical interventions. Instead, hypnosis works best when it works with traditional Western medicine to enhance the treatment program.

Hypnosis has demonstrated success in addressing a number of common issues that are encountered in many clinical settings. Hypnosis will not work for every patient, nor will it help every patient that has one of the issues listed below. It is more accurate to think of clinical hypnosis as one of a number of options available to the medical professional.

Consider the treatment of hypertension (high blood pressure): Although Norvasc is the most popular medication used to treat high blood pressure, it is not the right medication for every individual patient. There are dozens of high blood pressure drugs on the market, all of which have been proven to be effective at treating the condition. Hypnosis has proven to be effective in reducing blood pressure, but just like Norvasc, it may not be the right treatment for everyone.

This book summarizes the scientific evidence for hypnosis as a complementary therapy, including references and abstracts. This research has all been published in the peer-reviewed literature as cited in the MEDLINE database.

- **Stress management** Studies have shown hypnosis can reduce levels of stress hormones and have a positive impact on self-reported stress assessment scales.

- **Stress disorders (PTSD/ASD)** The primary benefit hypnosis offers in these disorders is as an additive to psychotherapy. Hypnosis can assist helping the patient address the fear that triggers the disorder condition.

- **Pre- and post-surgical complaints** Surgery can be psychologically traumatic, which can increase the chances of complications and prolong post-operative healing. Hypnosis can defuse the mental impact and improve the result of the surgery.

- **Lifestyle modification** A number of medical conditions can be positively impacted by changes in a patient's lifestyle. Hypnotists regularly assist their clients in making such changes, and can also assist medical professionals in managing patient care.

- **Pain reduction** Pain control has been well-studied and has shown promising results in multiple conditions. It has been shown capable of treating both chronic (long-term) and acute (short-term) pain.

- **Nausea control** Nausea is a common side effect of many treatments, or a symptom of a number of disease conditions. Hypnosis can assist in controlling nausea and related conditions.

- **Overcoming medical-related fears** Medical fears can keep a person from seeking necessary treatment or cause complications when they do. Hypnosis can help the patient manage such fears to improve treatment outcome.

- **Immune Function** Although many complementary or alternative treatments claim to "boost the immune system" as a catch-all benefit, hypnosis has been shown to improve immune function. Hypnosis can increase levels of white blood cells and immune-system antibodies.

In this book you will se a number of terms used to describe a related group of therapies. Terms like therapeutic suggestion, clinical suggestion, guided imagery, guided imagery with music, guided meditation, biofeedback, autogenic training, etc. are all related to therapies that utilize similar brain pathways. Therefore, this book demonstrates evidence for other mind modalities similar to hypnosis.

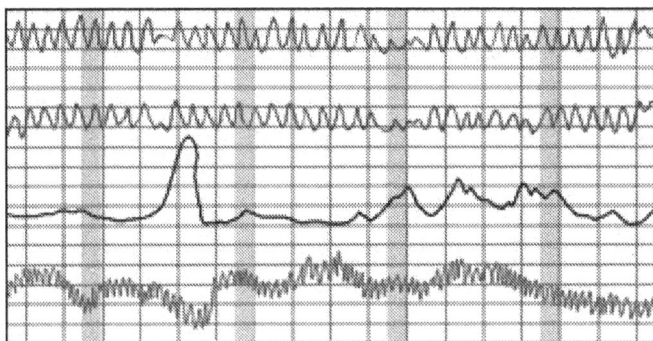

Stress Management

Summary

Medical professionals regularly deal with a range of stress-related illnesses. These conditions include cardiovascular disease, asthma, psoriasis, immune disorders, and psychological illness. The professional hypnotist can assist the management of these conditions by addressing the effects of stress on both a psychological and physiological level. Hypnosis can reduce the stress chemical levels and improve the patient/client's perception of their stress levels.

A group at the Portuguese Cancer Institute monitored the responses of subjects exposed to various emotional states while under hypnosis, and found correlations of the production of the hormones cortisol and prolactin to different emotions suggested by the researchers. When stressful emotions were suggested, cortisol levels were affected. When the subjects were suggested to think about breastfeeding, prolactin, a hormone associated with lactation, was affected. While prolactin is generally not considered a stress hormone, this does show the ability of hypnosis to affect the body's biochemical responses

Another group at the University of Miami found that a guided imagery/music relaxation technique decreased the level of cortisol, a hormone secreted by the adrenal gland in response to stress. Cortisol, as part of the body's fight-or-flight response, helps convert protein to carbohydrates and increase blood sugar, giving the body more energy to face danger and can be used as a biochemical indicator of the stress response.

Randomized clinical trials have also evaluated the effect of hypnosis on the mood of individuals using standardized stress assessment scales. Hypnosis caused reductions in state anxiety in both Multiple Sclerosis patients and nursing students. The students also reported fewer sick days while in the active treatment group.

It is well-established that emotional states affect the body's chemistry. Hypnosis can help doctors and patients positively impact the patient's overall health by reducing stress and therefore the negative impacts of stress on the body.

References

Cortisol, Prolactin, Growth Hormone And Neurovegetative Responses To Emotions Elicited During An Hypnoidal State

Sobrinho LG, Simões M, Barbosa L, Raposo JF, Pratas S, Fernandes PL, Santos MA. *Psychoneuroendocrinology*. 2003 Jan;28(1):1-17.

The present study describes the responses of cortisol, prolactin and growth hormone (GH) to emotions elicited during sessions in which an hypnoidal state was induced. The purpose of the study was to provide answers for the following questions: 1) Do sessions with an emotional content have more hormonal surges than baseline, relaxation-only, sessions? 2) Does the induction of a fantasy of pregnancy and nursing elicit a prolactin response? 3) Are there any associations between surges of different hormones? 4) Are hormonal responses related to the intensity, type, or mode of expression of the emotions? For this purpose, thirteen volunteers and twelve patients with minor emotional difficulties were studied during sessions under hypnosis. The period of observation lasted for about three hours. Heart rate (HR), skin conductance (SC) and vagal tone (VT) were monitored. Serum cortisol, prolactin and growth hormone were sampled every 15 minutes. The volunteers had three types of sessions- "blank", consisting of relaxation only (12 sessions), "breast feeding", in which a fantasy of pregnancy and breast feeding was induced (12 sessions) and "free associations" in which the subjects were encouraged to evoke experiences or feelings (17 sessions). The patients had only sessions of free associations (38 sessions). Sessions of free associations had more hormonal surges than "blank" and "breast feeding" sessions. This was true for cortisol (8/17 v.3/24; $p < 0.03$), prolactin (7/17 v. 3/24; $p < 0.05$) and GH (9/17 v. 4/24; $p < 0.02$). During the 55 sessions of free associations (volunteers plus patients) there were 32 surges of cortisol, 18 of prolactin and 28 of GH. Cortisol and prolactin surges were negatively correlated ($p < 0.03$). GH had no significant association with either cortisol or prolactin. Visible emotions were positively associated with GH surges ($p < 0.05$). but not with cortisol or prolactin. Cortisol surges were correlated positively with evocations of real events ($p < 0.01$) and negatively with evocations containing defensive elements ($p < 0.01$).

Cortisol, Prolactin, Growth Hormone ... (continued)

Cortisol correlated positively with shock and intimidation ($p < 0.02$) and negatively with rage ($p < 0.04$). The AUC of the cortisol peaks during shock and intimidation was significantly higher than that of the pool of all other cortisol peaks (12.4 micromol x min x l(-1) v. 7.1 micromol x min x l(-1); $p < 0.005$). Rage had a marginally significant positive association with prolactin surges ($p=0.07$). The distribution of GH surges did not show any significant association with types of emotions. The present study provides evidence that cortisol, prolactin and GH respond to psychological stress in humans. However, they are regulated differently from one another. Cortisol and prolactin surges appear to be alternative forms of response to specific emotions. GH surges depend on the intensity of the emotion, probably as a consequence of the associated muscular activity. The current paradigm of stress, implying corticotrophin-releasing hormone (CRH) as the initial step of a cascade of events, is insufficient to account for the diversity of hormonal changes observed in psychological stress in humans.

Effects Of Guided Imagery And Music (GIM) Therapy On Mood And Cortisol In Healthy Adults

McKinney CH, Antoni MH, Kumar M, Tims FC, McCabe PM. *Health Psychology*. 1997 Jul;16(4):390-400.

Healthy adults (N = 28) participated in a randomized trial of Bonny Method of Guided Imagery and Music (GIM; a depth approach to music psychotherapy) sessions on mood and cortisol. Participants in both GIM and wait-list control conditions completed the Profile of Mood States (POMS) and donated 15 cc of blood before and after the 13-week intervention period and again at a 6-week follow-up. Split-plot factorial and post hoc analyses demonstrated that after 6 biweekly sessions GIM participants reported significant decreases between pre- and postsession depression, fatigue, and total mood disturbance and had significant decreases in cortisol level by follow-up. Pretest to follow-up decrease in cortisol was significantly associated with decrease in mood disturbance. A short series of GIM sessions may positively affect mood and reduce cortisol levels in healthy adults. Such changes in hormonal regulation may have health implications for chronically stressed people.

The Effects Of Imagery On Attitudes And Moods In Multiple Sclerosis Patients

Maguire BL. *Alternative Therapies In Health And Medicine*. 1996 Sep;2(5):75-9.

OBJECTIVE: To determine the efficacy of imagery for influencing attitudes and moods in multiple sclerosis patients. DESIGN: Experimental pretest-posttest, control-group. SETTING: Outpatient group in Central Pennsylvania. PATIENTS: 33 patients with mean ages of 43.93 years in the imagery group and 46.33 years in the control group. All subjects previously were identified with multiple sclerosis. INTERVENTION: Control group subjects followed their typical medical protocol and completed pretest and posttest measures. Imagery group subjects completed pretest and posttest measures and participated in a six-session group process that included brief exposure to relaxation training and ongoing work with biologically oriented imagery. Relaxation training and imagery were practiced on a daily basis. Imagery group subjects also produced imagery drawings, which were assessed after the third and sixth sessions. MAIN OUTCOME MEASURES: Profile of Mood States, State-Trait Anxiety Inventory, Health Attribution Test, Imagery Assessment Tool, and Multiple Sclerosis Symptom Checklist. MAIN RESULTS: Imagery group subjects demonstrated significant reductions in state anxiety and significant alteration in their illness imagery because of feedback obtained during the study. CONCLUSION: Use of the relaxation/imagery protocol led to clinically significant reductions in state anxiety. Imagery may be assessed through drawings that allow for positive modification of the imagery material to increase its utility and power.

Autogenic Training To Reduce Anxiety In Nursing Students: Randomized Controlled Trial

Kanji N, White A, Ernst E. *Journal Of Advanced Nursing*. 2006 Mar;53(6):729-35.

AIM: This paper reports a study to determine the effectiveness of autogenic training in reducing anxiety in nursing students. BACKGROUND: Nursing is stressful, and nursing students also have the additional pressures and uncertainties shared with all academic students. Autogenic training is a relaxation technique consisting of six mental exercises and is aimed at relieving tension, anger and stress. Meta-analysis has found large effect sizes for autogenic trainings intervention comparisons, medium effect sizes against control groups, and no effects when compared with other psychological therapies. A controlled trial with 50 nursing students found that the number of certified days off sick was reduced by autogenic training compared with no treatment, and a second trial with only 18 students reported greater improvement in Trait Anxiety, but not State Anxiety, compared with untreated controls. METHODS: A randomized controlled trial with three parallel arms was completed in 1998 with 93 nursing students aged 19-49 years. The setting was a university college in the United Kingdom. The treatment group received eight weekly sessions of autogenic training, the attention control group received eight weekly sessions of laughter therapy, and the time control group received no intervention. The outcome measures were the State-Trait Anxiety Inventory, the Maslach Burnout Inventory, blood pressure and pulse rate completed at baseline, 2 months (end of treatment), and 5, 8, and 11 months from randomization. RESULTS: There was a statistically significantly greater reduction of State ($P<0.001$) and Trait ($P<0.001$) Anxiety in the autogenic training group than in both other groups immediately after treatment. There were no differences between the groups for the Maslach Burnout Inventory. The autogenic training group also showed statistically significantly greater reduction immediately after treatment in systolic ($P<0.01$) and diastolic ($P<0.05$) blood pressure, and pulse rate ($P<0.002$), than the other two groups. CONCLUSION. Autogenic training has at least a short-term effect in alleviating stress in nursing students.

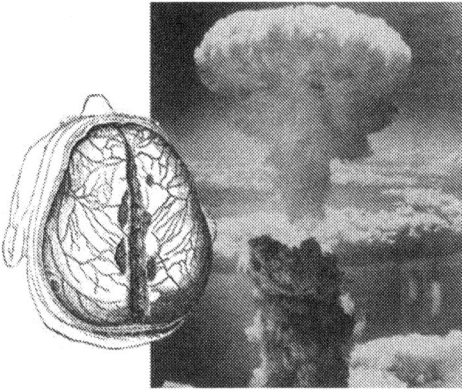

Stress Disorders (PTSD/ASD)

Summary

Specific stress disorders like Post-Traumatic Stress Disorder and Acute Stress Disorder have been recognized in various forms for many years. As far back as the Civil War, the term "Soldier's Heart" was coined to describe the mental changes brought on by exposure to psychologically shocking events. Clinical hypnosis in combination with psychotherapy can be an effective intervention in both disorders. Hypnosis can assist the patient address the fear that triggers the disorder condition.

ASD and PTSD have some similarities that allow us to consider them together. They are both conditions which occur in survivors of psychological trauma. That is, the people that suffer from these conditions have been through an experience that provoked intense fear or horror. They both produce "dissociative" symptoms like feelings of numbness, detachment, and even amnesia. Left unchecked, Acute Stress Disorder can in time produce Post-Traumatic Stress Disorder. Hypnosis can assist the treatment program to enhance the person's ability to function.

A group at the University of New South Wales in Australia investigated the usefulness of hypnosis with Cognitive Behavioral Therapy (CBT) in treating Acute Stress Disorder. Cognitive behavioral therapy is a psychological approach based on recent research into how people think and perceive the world around them. It attempts to identify and understand problems in terms of the relationship between thoughts, feelings and behavior. In a preliminary study, patients that were treated with CBT in conjunction with hypnosis had a lower chance of re-experiencing the traumatic event than patients treated with CBT alone. A long-term follow-up study supported this conclusion after 3 years.

Two Binghamton University professors reviewed the literature published on hypnosis and PTSD, finding that hypnosis can be a useful addition to psychological PTSD treatment. A case study from NYU highlights the way hypnosis can assist a psychological treatment program.

References

The Additive Benefit Of Hypnosis And Cognitive-Behavioral Therapy In Treating Acute Stress Disorder

Bryant RA, Moulds ML, Guthrie RM, Nixon RD. *Journal Of Consulting And Clinical Psychology*. 2005 Apr;73(2):334-40.

This research represents the first controlled treatment study of hypnosis and cognitive- behavioral therapy (CBT) of acute stress disorder (ASD). Civilian trauma survivors (N=87) who met criteria for ASD were randomly allocated to 6 sessions of CBT, CBT combined with hypnosis (CBT-hypnosis), or supportive counseling (SC). CBT comprised exposure, cognitive restructuring, and anxiety management. CBT-hypnosis comprised the CBT components with each imaginal exposure preceded by a hypnotic induction and suggestions to engage fully in the exposure. In terms of treatment completers (n=69), fewer participants in the CBT and CBT-hypnosis groups met criteria for posttraumatic stress disorder at posttreatment and 6-month follow-up than those in the SC group. CBT-hypnosis resulted in greater reduction in reexperiencing symptoms at posttreatment than CBT. These findings suggest that hypnosis may have use in facilitating the treatment effects of CBT for posttraumatic stress.

Hypnotherapy And Cognitive Behaviour Therapy Of Acute Stress Disorder: A 3-Year Follow-Up

Bryant RA, Moulds ML, Nixon RD, Mastrodomenico J, Felmingham K, Hopwood S. *Behaviour Research And Therapy*. 2006 Sep;44(9):1331-5.

The long-term benefits of cognitive behaviour therapy (CBT) for trauma survivors with acute stress disorder were investigated by assessing patients 3 years after treatment. Civilian trauma survivors (n=87) were randomly allocated to six sessions of CBT, CBT combined with hypnosis, or supportive counselling (SC), 69 completed treatment, and 53 were assessed 2 years post-treatment for post-traumatic stress disorder (PTSD) with the Clinician-Administered PTSD Scale. In terms of treatment completers, 2 CBT patients (10%), 4 CBT/hypnosis patients (22%), and 10 SC patients (63%) met PTSD criteria at 2-years follow-up. Intent-to-treat analyses indicated that 12

Hypnotherapy And Cognitive Behaviour Therapy... (continued)

CBT patients (36%), 14 CBT/hypnosis patients (46%), and 16 SC patients (67%) met PTSD criteria at 2-year follow-up. Patients who received CBT and CBT/hypnosis reported less re-experiencing and less avoidance symptoms than patients who received SC. These findings point to the long-term benefits of early provision of CBT in the initial month after trauma.

Hypnosis And The Treatment Of Posttraumatic Conditions: An Evidence-Based Approach

Lynn SJ, Cardeña E. *The International Journal of Clinical and Experimental Hypnosis*. 2007 Apr;55(2):167-88.

This article reviews the evidence for the use of hypnosis in the treatment of posttraumatic conditions including posttraumatic stress disorder and acute stress disorder. The review focuses on empirically supported principles and practices and suggests that hypnosis can be a useful adjunctive procedure in the treatment of posttraumatic conditions. Cognitive-behavioral and exposure-based interventions, which have the greatest empirical support, are highlighted, and an illustrative case study is presented.

Hypnotic Imagery Rehearsal In The Treatment Of Nightmares: A Case Report

Donatone B. *The American Journal of Clinical Hypnosis*. 2006 Oct;49(2):123-7.

This case report discusses a patient who experienced frequent nightmares and chronic low-level anxiety during his 3 1/2 year imprisonment. He developed post traumatic stress disorder (PTSD), in part because he adamantly insisted that he had been wrongfully incarcerated. The literature supports the use of hypnotic imagery rehearsal for treating nightmares that stem from PTSD. Due to the patient's distrust of others and trauma history, it was uncertain whether hypnotic intervention would be effective. It is of note, there is no indication in the literature that hypnosis has been used with people on parole, let alone individuals who believe they were wrongly accused of committing a crime.

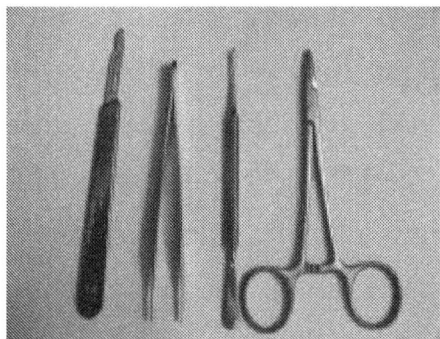

Pre- and Post-Surgical Complaints

Summary

Surgery, even minor or elective surgery, is usually physically traumatic. It can also cause mental trauma to the patient. The surgical environment is intimidating and the surgical experience is alien. Mental distress can increase complications, reduce patient compliance, increase perceived pain, and generally lengthen the recovery process. By addressing patient fears and distress, clinical hypnosis can improve the surgical outcome.

A group from University College, London used hypnosis to "rehearse" upcoming abdominal surgery. Patients that had this preparation had less pain and also felt better able to cope with the pain they did experience. Furthermore, their post-operative cortisol levels were also lower.

The Cleveland Clinic has a Guided Imagery Program which regularly uses hypnosis for surgical patients. As in the University College study, patients that listened to guided imagery tapes before, during, and after their surgeries had reduced pain and anxiety, and required lower amounts narcotic pain medication. The Rainbow Babies and Children's Hospital, also in Cleveland, showed similar results in preparing children for surgery. Studies performed at Guy's Hospital in London, England and in Sweden back up these results.

In addition to the pain and anxiety reduction, the British study also showed hypnosis can result in shorter post-surgical hospital stays, which is a major concern in today's hospitals. Both managed care companies and hospital administrations are actively trying to shorten hospital stays as much as possible, and hypnosis is a tool that can assist that goal.

References

Preoperative Rehearsal Of Active Coping Imagery Influences Subjective And Hormonal Responses To Abdominal Surgery

Manyande A, Berg S, Gettins D, Stanford SC, Mazhero S, Marks DF, Salmon P.
Psychosomatic Medicine. 1995 Mar-Apr;57(2):177-82.

Existing evidence suggests that preoperative psychological preparation that is designed to reduce anxiety may sensitize cortisol and adrenaline responses to surgery. In a controlled trial of abdominal surgery patients, we therefore tested the effects of a preoperative preparation that used guided imagery, not to reduce anxiety, but to increase patients' feelings of being able to cope with surgical stress; 26 imagery patients were compared with 25 controls who received, instead, background information about the hospital. State-anxiety was similar in each group, but imagery patients experienced less postoperative pain than did the controls, were less distressed by it, felt that they coped with it better, and requested less analgesia. Hormone levels measured in peripheral venous blood did not differ on the afternoon of admission, before preparation. Cortisol levels were, however, lower in imagery patients than in controls immediately before and after surgery. Noradrenaline levels were greater on these occasions in imagery patients than controls. The results are interpreted in relation to two theories. One states that preoperative "worry" reduces surgical stress. The other concerns the influence of active and passive coping on endocrine responses to stress.

Guided Imagery As A Coping Strategy For Perioperative Patients

Tusek D, Church JM, Fazio VW. *AORN Journal*. 1997 Oct;66(4):644-9.

Patients who undergo surgery usually experience fear and apprehension about their surgical procedures. Guided imagery is a simple, low-cost therapeutic tool that can help counteract surgical patients' fear and anxiety. The authors randomly assigned 130 patients undergoing elective colorectal surgical procedures into two groups. Members of one group received routine perioperative care. Members of the other group listened to guided imagery tapes for three days before their surgical procedures, during anesthesia induction, intraoperatively, in the postanesthesia care unit, and for six days after surgery. The authors measured patients' anxiety levels, pain perceptions, and narcotic medication requirements. The patients in the guided imagery group experienced considerably less preoperative and postoperative anxiety and pain, and they required almost 50% less narcotic medications after their surgical procedures than patients in the control group.

The Effects Of Hypnosis/Guided Imagery On The Postoperative Course Of Children

Lambert SA. *Journal Of Developmental And Behavioral Pediatrics*. 1996 Oct;17(5): 307-10.

Hypnosis, guided imagery, and relaxation have been shown to improve the postoperative course of adult surgical patients. Children have successfully used hypnosis/guided imagery to significantly reduce the pain associated with invasive procedures and to improve selected medical conditions. The purpose of this study was to examine the effect of hypnosis/guided imagery on the postoperative course of pediatric surgical patients. Fifty-two children (matched for sex, age, and diagnosis) were randomly assigned to an experimental or control group. The experimental group was taught guided imagery by the investigator. Practice of the imagery technique included suggestions for a favorable postoperative course. Significantly lower postoperative pain ratings and shorter hospital stays occurred for children in the

The Effects Of Hypnosis/Guided Imagery... (continued)

experimental group. State anxiety was decreased for the guided imagery group and increased postoperatively for the control group. This study demonstrates the positive effects of hypnosis/guided imagery for the pediatric surgical patient.

Improved Recovery And Reduced Postoperative Stay After Therapeutic Suggestions During General Anaesthesia

Evans C, Richardson PH. *Lancet*. 1988 Aug 27;2(8609):491-3.

The clinical value of therapeutic suggestions during general anaesthesia was assessed in a double-blind randomised placebo-controlled study. 39 unselected patients were allocated to suggestion (n = 19) or control (n = 20) groups who were played either recorded therapeutic suggestions or a blank tape, respectively, during hysterectomy. The patients in the suggestion group spent significantly less time in hospital after surgery, suffered from a significantly shorter period of pyrexia, and were generally rated by nurses as having made a better than expected recovery. Patients in the suggestion group, unlike those in the control group, guessed accurately that they had been played an instruction tape.

Improved Recovery After Music And Therapeutic Suggestions During General Anaesthesia: A Double-Blind Randomised Controlled Trial

Nilsson U, Rawal N, Unestähl LE, Zetterberg C, Unosson M. *Acta Anaesthesiologica Scandinavica*. 2001 Aug;45(7):812-7.

PURPOSE: This study was designed to determine whether music or music in combination with therapeutic suggestions in the intra-operative period under general anaesthesia could improve the recovery of hysterectomy patients. METHODS: In a double-blind randomised clinical investigation, 90 patients who underwent hysterectomy under general anaesthesia were intra-operatively exposed to music, music in combination with therapeutic suggestion or operation room sounds. The anaesthesia was standardised. Postoperative analgesia was

19

Improved Recovery After Music And Therapeutic Suggestions...(continued)

provided by a patient-controlled analgesia (PCA). The pain scores were recorded by means of a visual analogue scale. Nausea, emesis, bowel function, fatigue, well-being and duration of hospital stay were studied as outcome variables. RESULTS: On the day of surgery, patients exposed to music in combination with therapeutic suggestions required less rescue analgesic compared with the controls. Patients in the music group experienced more effective analgesia the first day after surgery and could be mobilised earlier after the operation. At discharge from the hospital patients in the music and music combined with therapeutic suggestion group were less fatigued compared to the controls. No differences were noted in nausea, emesis, bowel function, well-being or length of hospital stay between the groups. CONCLUSION: This double-blind study has demonstrated that intra-operative music and music in combination with therapeutic suggestions may have some beneficial effects on postoperative recovery after hysterectomy. Further controlled studies are necessary to confirm our results.

Lifestyle Modifications

Summary

For most hypnotists, helping clients change their lifestyle by for example, stopping smoking, losing weight, or stopping drinking, are part of their bread and butter. Many medical conditions are negatively impacted by these lifestyle factors. This includes the impact of smoking on cancer or heart disease, drinking on liver disease and alcoholic cirrhosis, and weight on a host of conditions. With increasing frequency, medical professionals are likely to address these issues as part of their total care for their patients. The hypnotist can be an effective ally in helping the medical professional manage these issues.

Smoking

A recent review by psychologists at Harvard Medical School on hypnosis and smoking showed that hypnosis can contribute significantly to smoking cessation programs. Example studies that justify this conclusion are:

- The American Lung Association of Ohio offers single-session hypnosis as an option for its smoking cessation programs. In a study of this program by Ohio State University, 22% of participants reported success, higher than the percentage that used anti-smoking medication (20%).

- A study of single-session hypnosis to quit smoking at the University of Scranton showed significantly more nonsmoking in the treated group than either a control group or a placebo group. The reduction in smoking was seen not only immediately after treatment, but as long as 48 weeks later.

- A group at the Morriston Hospital in Wales exposed female smokers to suggestion during anesthesia and those that heard an "active" tape containing post-hypnotic suggestions were more likely to have stopped smoking after one month.

Substance Abuse

Hypnosis can also positively impact treatment of other addictive behaviors, such as alcoholism or drug abuse. One clinician reports that 77% of patients in his practice treated with hypnosis in conjunction with counseling were not drinking one year later. Cocaine abusers in another study from Yale University showed lower levels of cravings after treatment with a relaxation exercise.

Weight Loss

Hypnosis is not a guaranteed method to lose weight, but it has demonstrated the ability to enhance a total program for weight loss. A group at Churchill Hospital in Oxford, England found adding hypnosis to a dietary advice program resulted in significantly greater weight loss. Adding hypnosis to a behavioral therapy weight loss program also showed significantly greater weight loss, as long as two years later. A researcher at the University of Connecticut performed a large-scale synthesis of published research on weight loss and hypnosis, finding that the programs that added hypnosis increased weight loss.

References

Smoking

Hypnosis, Behavioral Theory, And Smoking Cessation

Covino NA, Bottari M. *Journal of Dental Education.* 2001 Apr;65(4):340-7

Although nicotine replacement and other pharmacological treatments head the list of popular interventions for smoking cessation, approaches based on psychology can also assist smokers. Hypnosis, suggestion, and behavior therapies have been offered to patients and studied experimentally for several decades. Although no single psychological approach has been found to be superior to others, psychological interventions contribute significantly to successful treatment outcome in smoking cessation. This article describes common hypnotic and behavioral approaches to smoking cessation and critically reviews some of the findings from clinical and experimental research studies. The authors also offer suggestions regarding treatment and future research.

Descriptive Outcomes Of The American Lung Association Of Ohio Hypnotherapy Smoking Cessation Program

Ahijevych K, Yerardi R, Nedilsky N. *International Journal of Clinical and Experimental Hypnosis.* 2000 Oct;48(4):374-87.

Hypnosis smoking cessation treatment is one type of program available to smokers. This paper describes a large randomly selected sample from such a program, which has not been previously reported. During 1997, 2,810 smokers participated in single-session, group hypnotherapy smoking cessation programs sponsored by the American Lung Association of Ohio. A randomly selected sample of 452 participants completed telephone interviews 5 to 15 months after attending a treatment session. Twenty-two percent of participants (n = 101) reported not smoking during the month prior to the interview. Use of other smoking cessation strategies since the treatment session were assessed. Interestingly, only 20% of participants who used

Descriptive Outcomes Of The American Lung Association... (continued)

pharmaceutical products to assist with smoking cessation took them for the recommended treatment duration. Hypnotherapy smoking cessation treatment offers an alternative cessation method, which may meet the unique needs of certain individuals

Use Of Single Session Hypnosis For Smoking Cessation

Williams JM, Hall DW. *Addictive Behaviors.* 1988;13(2):205-8.

Twenty of sixty volunteers for smoking cessation were assigned to single-session hypnosis, 20 to a placebo control condition, and 20 to a no-treatment control condition. The single-session hypnosis group smoked significantly less cigarettes and were significantly more abstinent than a placebo control group and a no treatment control group at posttest, and 4-week, 12-week, 24-week and 48-week follow-ups.

Reducing Smoking. The Effect Of Suggestion During General Anaesthesia On Postoperative Smoking Habits

Hughes JA, Sanders LD, Dunne JA, Tarpey J, Vickers MD. *Anaesthesia.* 1994 Feb;49(2):126-8.

In a double-blind randomised trial, 122 female smokers undergoing elective surgery were allocated to receive one of two prerecorded messages while fully anaesthetised. The active message was designed to encourage them to give up smoking whilst the control message was the same voice counting numbers. No patient could recall hearing the tape. Patients were asked about their postoperative smoking behaviour one month later. Significantly more of those who had received the active tape had stopped or reduced their smoking (p < 0.01). This would suggest a level of preconscious processing of information.

Substance Abuse

Intensive Therapy: Utilizing Hypnosis In The Treatment Of Substance Abuse Disorders

Potter G. *The American Journal Of Clinical Hypnosis*. 2004 Jul;47(1):21-8.

Hypnosis was once a viable treatment approach for addictions. Then, due to hypnosis being used for entertainment purposes many professionals lost confidence in it. However, it has now started to make a comeback in the treatment of substance abuse. The approach described here, using hypnosis for treatment, is borrowed from studies effectively treating alcoholism by using intensive daily sessions. Combining the more intense treatment of 20 daily sessions with hypnosis is a successful method to treat addictions. The treatment has been used with 18 clients over the last 7 years and has shown a 77 percent success rate for at least a 1-year follow-up.

Cue-Elicited Cocaine Craving And Autogenic Relaxation. Association With Treatment Outcome

Margolin A, Avants SK, Kosten TR. *Journal Of Substance Abuse Treatment*. 1994 Nov-Dec;11(6):549-52.

Prior to entering a pharmacotherapy trial for the treatment of cocaine dependence, 19 patients participated in a pretreatment cue-reactivity protocol that concluded with a relaxation exercise. Measures included self-reported craving and skin conductance level (SCL). Post hoc exploratory analyses suggest that neither craving nor change in SCL in response to cocaine cues differentiated patients who subsequently achieved abstinence from those who did not. Craving following the relaxation procedure did differentiate the two groups: patients who subsequently initiated abstinence in treatment reported a reduction in cue-elicited craving to below baseline levels; craving reported by patients who did not successfully complete treatment remained elevated.

Weight Loss

Controlled Trial Of Hypnotherapy For Weight Loss In Patients With Obstructive Sleep Apnoea

Stradling J, Roberts D, Wilson A, Lovelock F. *International Journal Of Obesity And Related Metabolic Disorders*. 1998 Mar;22(3):278-81.

OBJECTIVE: To assess if hypnotherapy assists attempts at weight loss. DESIGN: Randomised, controlled, parallel study of two forms of hypnotherapy (directed at stress reduction or energy intake reduction), vs dietary advice alone in 60 obese patients with obstructive sleep apnoea on nasal continuous positive airway pressure treatment. SETTING: National Health Service hospital in the UK. MEASURES: Weight lost at 1, 3, 6, 9, 12, 15 and 18 months after dietary advice and hypnotherapy, as a percentage of original body weight. RESULTS: All three groups lost 2-3% of their body weight at three months. At 18 months only the hypnotherapy group (with stress reduction) still showed a significant (P < 0.02), but small (3.8 kg), mean weight loss compared to baseline. Analysed over the whole time period the hypnotherapy group with stress reduction achieved significantly more weight loss than the other two treatment arms (P < 0.003), which were not significantly different from each other. CONCLUSIONS: This controlled trial on the use of hypnotherapy, as an adjunct to dietary advice in producing weight loss, has produced a statistically significant result in favour of hypnotherapy. However, the benefits were small and clinically insignificant. More intensive hypnotherapy might of course have been more successful, and perhaps the results of the trial are sufficiently encouraging to pursue this approach further.

Effectiveness Of Hypnosis As An Adjunct To Behavioral Weight Management

Bolocofsky DN, Spinler D, Coulthard-Morris L. *Journal of Clinical Psychology*. 1985 Jan;41(1):35-41.

This study examined the effect of adding hypnosis to a behavioral weight-management program on short- and long-term weight change. One hundred nine subjects, who ranged in age from 17 to 67, completed a behavioral treatment either with or without the addition

Effectiveness Of Hypnosis As An Adjunct...(continued)

of hypnosis. At the end of the 9-week program, both interventions resulted in significant weight reduction. However, at the 8-month and 2-year follow-ups, the hypnosis clients showed significant additional weight loss, while those in the behavioral treatment exhibited little further change. More of the subjects who used hypnosis also achieved and maintained their personal weight goals. The utility of employing hypnosis as an adjunct to a behavioral weight-management program is discussed.

Hypnotic Enhancement Of Cognitive-Behavioral Weight Loss Treatments--Another Meta-Reanalysis

Kirsch I. *Journal Of Consulting And Clinical Psychology*. 1996 Jun;64(3):517-9.

In a 3rd meta-analysis of the effect of adding hypnosis to cognitive-behavioral treatments for weight reduction, additional data were obtained from authors of 2 studies, and computational inaccuracies in both previous meta-analyses were corrected. Averaged across posttreatment and follow-up assessment periods, the mean weight losswas 6.00 lbs. (2.72 kg) without hypnosis and 11.83 lbs. (5.37 kg) with hypnosis. The mean effect size of this difference was 0.66 SD. At the last assessment period, the mean weight loss was 6.03 lbs. (2.74 kg) without hypnosis and 14.88 lbs. (6.75 kg) with hypnosis. The effect size for this difference was 0.98 SD. Correlational analyses indicated that the benefits of hypnosis increased substantially over time (r = .74).

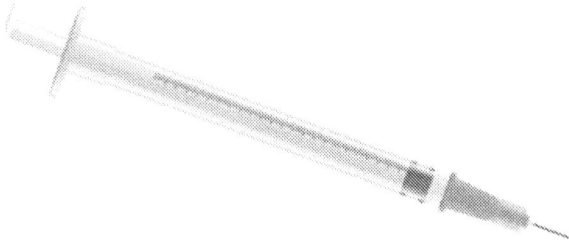

Pain Reduction

Summary

Pain control is probably the application of hypnosis in the clinical setting that has been both best-studied and has seen the greatest acceptance by medical professionals. By adding hypnosis to a patient's treatment, medical professionals can enhance patient well-being.

When hypnosis is used to help treat chronic pain, it has been shown to have a definite and significant impact. Although researchers at the University of Washington had some concerns that the research on the subject has not eliminated the placebo effect, a group at Texas A&M University did not have this concern, and both groups showed significant decreases in pain.

Hypnosis has also shown usefulness in reducing pain during various treatment procedures. Examples of such use include:

- Reducing pain during bone marrow transplant procedures at the Fred Hutchinson Cancer Research Center in Seattle

- Reducing pain and anxiety during wisdom-tooth surgery at the University of Iowa

- Reducing the amount of pain-control drugs and procedure interruptions required during interventional radiology procedures at the Department of Veterans Affairs Medical Center in Palo Alto, California

Pain control that uses hypnosis also has other, sometimes unexpected, benefits. A group at the University of Washington found that patients who had hypnotic treatments were more satisfied both with their treatment and their ability to cope with the pain they did have, even if the reduction in pain levels was not substantial. Increasing patient satisfaction is always beneficial to both the doctor and patient.

References

Hypnotic Treatment Of Chronic Pain

Jensen M, Patterson DR. *Behavioral Medicine*. 2006 Feb;29(1):95-124. Epub 2006 Jan 11.

This article reviews controlled trials of hypnotic treatment for chronic pain in terms of: (1) analyses comparing the effects of hypnotic treatment to six types of control conditions; (2) component analyses; and (3) predictor analyses. The findings indicate that hypnotic analgesia produces significantly greater decreases in pain relative to no-treatment and to some non-hypnotic interventions such as medication management, physical therapy, and education/advice. However, the effects of self-hypnosis training on chronic pain tend to be similar, on average, to progressive muscle relaxation and autogenic training, both of which often include hypnotic-like suggestions. None of the published studies have compared hypnosis to an equally credible placebo or minimally effective pain treatment, therefore conclusions cannot yet be made about whether hypnotic analgesia treatment is specifically effective over and above its effects on patient expectancy. Component analyses indicate that labeling versus not labeling hypnosis treatment as hypnosis, or including versus not including hand-warming suggestions, have relatively little short-term impact on outcome, although the hypnosis label may have a long-term benefit. Predictor analyses suggest that global hypnotic responsivity and ability to experience vivid images are associated with treatment outcome in hypnosis, progressive relaxation, and autogenic training treatments. The paper concludes with a discussion of the implications of the findings for future hypnosis research and for the clinical applications of hypnotic analgesia.

Hypnotherapy For The Management Of Chronic Pain

Elkins G, Jensen MP, Patterson DR. *The International Journal of Clinical and Experimental Hypnosis.* 2007 Jul;55(3):275-87.

This article reviews controlled prospective trials of hypnosis for the treatment of chronic pain. Thirteen studies, excluding studies of headaches, were identified that compared outcomes from hypnosis for the treatment of chronic pain to either baseline data or a control condition. The findings indicate that hypnosis interventions consistently produce significant decreases in pain associated with a variety of chronic-pain problems. Also, hypnosis was generally found to be more effective than nonhypnotic interventions such as attention, physical therapy, and education. Most of the hypnosis interventions for chronic pain include instructions in self-hypnosis. However, there is a lack of standardization of the hypnotic interventions examined in clinical trials, and the number of patients enrolled in the studies has tended to be low and lacking long-term follow-up. Implications of the findings for future clinical research and applications are discussed.

Relaxation And Imagery And Cognitive-Behavioral Training Reduce Pain During Cancer Treatment: A Controlled Clinical Trial

Syrjala KL, Donaldson GW, Davis MW, Kippes ME, Carr JE. *Pain.* 1995 Nov;63(2): 189-98.

Few controlled clinical trials of psychological interventions for cancer pain relief exist in spite of frequent support for their importance as adjuncts to medical treatment. This study compared oral mucositis pain levels in 4 groups of cancer patients receiving bone marrow transplants (BMT): (1) treatment as usual control, (2) therapist support, (3) relaxation and imagery training, and (4) training in a package of cognitive-behavioral coping skills which included relaxation and imagery. A total of 94 patients completed the study which involved two training sessions prior to treatment and twice a week 'booster' sessions during the first 5 weeks of treatment. Results confirmed our hypothesis that patients who received either relaxation and imagery alone or patients who received the package of cognitive-behavioral coping skills would report less pain than patients in the other 2 groups.

Relaxation And Imagery And Cognitive-Behavioral... (continued)

The hypothesis that the cognitive-behavioral skills package would have an additive effect beyond relaxation and imagery alone was not confirmed. Average visual analogue scale (VAS) report of pain within the therapist support group was not significantly lower than the control group (P = 0.103) nor significantly higher than the training groups. Patient reports of relative helpfulness of the interventions for managing pain and nausea matched the results of VAS reports. From these results, we conclude that relaxation and imagery training reduces cancer treatment-related pain; adding cognitive-behavioral skills to the relaxation with imagery does not, on average, further improve pain relief.

Tape-Recorded Hypnosis Instructions As Adjuvant In The Care Of Patients Scheduled For Third Molar Surgery

Ghoneim MM, Block RI, Sarasin DS, Davis CS, Marchman JN. *Anesthesia and Analgesia*. 2000 Jan;90(1):64-8.

As medical costs continue to escalate, there is willingness to consider the role played by nontraditional factors in health. We investigated the usefulness of tape-recorded hypnosis instruction on perioperative outcome in surgical patients in a prospective, randomized, and partially blinded study. Sixty patients scheduled for third molar surgery were studied. Patients were allocated to either an experimental group (E) or a control group (C). Group E received an audio tape to listen to daily for the immediate preoperative week, which guided the patients through a hypnotic induction and included suggestions on enhancement of perioperative well-being. Group C did not receive any tapes. The same surgeon administered local anesthesia and a standard regimen of sedation and performed the operation for all patients. The following variables were assessed 1 wk before surgery, immediately before and after surgery, and for 3 days after surgery by the indicated measurements: State anxiety by a Spielberger scale; nausea and pain by visual analog scales; number of tablets of the analgesics that were used; number of episodes of vomiting; and complications. In addition, the surgeon's assessment of ease of surgery was recorded. Two

Tape-Recorded Hypnosis Instructions As Adjuvant... (continued)

variables showed differences between the groups. First, Group C exhibited a mean increase of 11.7 points on the Spielberger scale from the screening to the presurgery period, while Group E showed only a mean increase of 5.5 points during the same period, P = 0.01. Second, the mean number of vomiting episodes was more in Group E, 1.3, than in Group C, 0.3, P = 0.02. In conclusion, anxiety was reduced before surgery by means of an audio tape containing hypnotic instructions; however, for no apparent reason, there was also an increase in the incidence of vomiting. IMPLICATIONS: We administered hypnosis instructions to patients before third molar surgery. Anxiety was reduced, but there was an increase in the incidence of vomiting. Although an easy and cost-effective method, the value of this approach remains to be established.

Self-Hypnotic Relaxation During Interventional Radiological Procedures: Effects On Pain Perception And Intravenous Drug Use

Lang EV, Joyce JS, Spiegel D, Hamilton D, Lee KK. *The International Journal Of Clinical And Experimental Hypnosis*. 1996 Apr;44(2):106-19.

The authors evaluated whether self-hypnotic relaxation can reduce the need for intravenous conscious sedation during interventional radiological procedures. Sixteen patients were randomized to a test group, and 14 patients were randomized to a control group. All had patient-controlled analgesia. Test patients additionally had self-hypnotic relaxation and underwent a Hypnotic Induction Profile test. Compared to controls, test patients used less drugs (0.28 vs. 2.01 drug units; p < .01) and reported less pain (median pain rating 2 vs. 5 on a 0-10 scale; p < .01). Significantly more control patients exhibited oxygen desaturation and/or needed interruptions of their procedures for hemodynamic instability. Benefit did not correlate with hypnotizability. Self-hypnotic relaxation can reduce drug use and improve procedural safety.

Satisfaction With, And The Beneficial Side Effects Of, Hypnotic Analgesia

Jensen MP, McArthur KD, Barber J, Hanley MA, Engel JM, Romano JM, Cardenas DD, Kraft GH, Hoffman AJ, Patterson DR. *The International Journal of Clinical and Experimental Hypnosis.* 2006 Oct;54(4):432-47.

Case study research suggests that hypnosis treatment may provide benefits that are not necessarily the target of specific suggestions. To better understand satisfaction with and the beneficial "side effects" of hypnosis treatment, questions inquiring about treatment satisfaction and treatment benefits were administered to a group of 30 patients with chronic pain who had participated in a case series of hypnotic analgesia treatment. The results confirmed the authors' clinical experience and showed that most participants reported satisfaction with hypnosis treatment even when the targeted symptom (in this case, pain intensity) did not decrease substantially. Study participants also reported a variety of both symptom-related and nonsymptom-related benefits from hypnosis treatment, including decreased pain, increased perceived control over pain, increased sense of relaxation and well-being, and decreased perceived stress, although no single benefit was noted by a majority of participants.

Nausea Control

Summary

Medical professionals commonly face nausea issues in caring for patients. Nausea can be caused by drugs such as pain medications or antibiotics, by treatments such as chemotherapy, or as a side-effect of many anesthetics. Hypnosis can reduce nausea and therefore increase patient compliance.

A comprehensive review of published studies on the use of hypnosis to reduce nausea during chemotherapy for cancer patients done at the University of Plymouth in England revealed a large benefit from hypnosis. Hypnosis was at least as effective as psychological approaches in reducing nausea.

Post-surgical nausea and vomiting is also a major medical concern. Hypnosis has also demonstrated clinical value in addressing this issue. The Karolinska Institute in Sweden found that hypnosis before surgery drastically reduced postoperative vomiting, and also reduced the need for post-surgical pain medication.

Other groups have experimented with suggestions during the surgery while the patients are under anesthesia. Groups at Poole General Hospital in Dorset, England, and at the University of Ulm in Germany both showed significant reductions in nausea and vomiting using this technique.

References

Hypnosis For Nausea And Vomiting In Cancer Chemotherapy: A Systematic Review Of The Research Evidence

Richardson J, Smith JE, McCall G, Richardson A, Pilkington K, Kirsch I. *European Journal Of Cancer Care.* 2007 Sep;16(5):402-12.

To systematically review the research evidence on the effectiveness of hypnosis for cancer chemotherapy-induced nausea and vomiting (CINV). A comprehensive search of major biomedical databases including MEDLINE, EMBASE, ClNAHL, PsycINFO and the Cochrane Library was conducted. Specialist complementary and alternative medicine databases were searched and efforts were made to identify unpublished and ongoing research. Citations were included from the databases' inception to March 2005. Randomized controlled trials (RCTs) were appraised and meta-analysis undertaken. Clinical commentaries were obtained. Six RCTs evaluating the effectiveness of hypnosis in CINV were found. In five of these studies the participants were children. Studies report positive results including statistically significant reductions in anticipatory and CINV. Meta-analysis revealed a large effect size of hypnotic treatment when compared with treatment as usual, and the effect was at least as large as that of cognitive-behavioural therapy. Meta-analysis has demonstrated that hypnosis could be a clinically valuable intervention for anticipatory and CINV in children with cancer. Further research into the effectiveness, acceptance and feasibility of hypnosis in CINV, particularly in adults, is suggested. Future studies should assess suggestibility and provide full details of the hypnotic intervention.

Preoperative Hypnosis Reduces Postoperative Vomiting After Surgery Of The Breasts. A Prospective, Randomized And Blinded Study

Enqvist B, Björklund C, Engman M, Jakobsson J. *Acta Anaesthesiologica Scandinavica*. 1997 Sep;41(8):1028-32.

BACKGROUND: Postoperative nausea and vomiting (PONV) after general anesthesia and surgery may have an incidence as high as 70% irrespective of antiemetic drug therapy. The use of preoperative hypnosis and mental preparation by means of an audio tape was investigated in the prophylaxis of nausea and vomiting before elective breast reduction surgery. Similar interventions have not been found in the literature. METHODS: Fifty women were randomized to a control group or a hypnosis group; the latter listened to an audio tape daily 4-6 days prior to surgery. A hypnotic induction was followed by suggestions as to how to relax and experience states incompatible with nausea and vomiting postoperatively (e.g. thirst and hunger). There was a training part on the tape where the patients were asked to rehearse their own model for stress reduction. Premedication and anesthetic procedures were standardized. RESULTS: Patients in the hypnosis group had significantly less vomiting, 39% compared to 68% in the control group, less nausea and less need of analgesics postoperatively. CONCLUSIONS: Preoperative relaxation and/or hypnotic techniques in breast surgery contribute to a reduction of both PONV and postoperative analgesic requirements.

The Incidence And Severity Of Postoperative Nausea And Vomiting In Patients Exposed To Positive Intra-Operative Suggestions

Williams AR, Hind M, Sweeney BP, Fisher R. *Anaesthesia*. 1994 Apr;49(4):340-2.

In a double-blind study, the effects of positive intra-operative suggestions on the incidence and severity of postoperative nausea and vomiting were studied in 60 patients randomly selected to undergo routine major gynaecological surgery. Patients who received positive suggestions suffered significantly less nausea and vomiting in the 24 h after surgery.

Therapeutic Suggestions Given During Neurolept-Anaesthesia Decrease Post-Operative Nausea And Vomiting

Eberhart LH, Döring HJ, Holzrichter P, Roscher R, Seeling W. *European Journal Of Anaesthesiology*. 1998 Jul;15(4):446-52.

A double-blind randomized study was performed in 100 patients undergoing thyroidectomy to evaluate the effect of positive therapeutic suggestions made during neurolept-anaesthesia. The classic droperidol-fentanyl-N2O technique was used as these drugs preserve the neurophysiological functions required to process the information in the therapeutic suggestions given during general anaesthesia. Patients in the suggestion group heard positive non-affirmative suggestions during the whole operation. An autoreverse tape player was used. The control group listened to an empty tape. Both groups were comparable with respect to demographic variables, anaesthetic technique, drug dosage, duration of anaesthesia and surgery. Patients in the suggestion group suffered significantly less from post-operative nausea or vomiting (suggestion: 47.2% vs. control: 85.7%) and required less anti-emetic treatment (suggestion: 30.6% vs. control: 68.6%). We conclude that therapeutic suggestions heard during neurolept-anaesthesia are processed and decrease post-operative nausea and vomiting in patients after thyroidectomy.

Overcoming Medical-Related Fears

Summary

A patient that has a fear of doctors or other medical-related fears may never even see a doctor or nurse. Even when a person with such fears does seek treatment, these fears can cause major complications. For example, a surgeon of the authors' acquaintance had a patient that feared dying during upcoming necessary surgery. While undergoing the surgery, the patient went into cardiac arrest for no apparent reason. Although the patient was resuscitated, the incident shows the power of negative fears.

A case reported by the Central Middlesex Hospital in London, England is a good example of a phobia that could have prevented vital treatment. A pregnant patient was suspected to have cephalopelvic disproportion, a condition that arises when the infant's head is too large to pass through the mother's pelvis. The safest method of delivering such an infant is through cesarean section (C-section), which has to be performed in a hospital environment. Five 30-minute hypnosis treatments allowed this patient to overcome her hospital phobia, which otherwise could have resulted in severe injury to the infant.

Phobias can also prevent or complicate dental treatment. At the University of Pennsylvania, children who had previously had violent emotional reactions to dental treatment were given hypnosis and sedation. The combination therapy was very effective at allowing treatment.

Dental phobia can also cause adults to forgo necessary dental treatment. The Royal Dental College of Denmark compared hypnosis as a treatment for dental anxiety to two other therapies, and found after 3 years, 54.5% of the patients treated with hypnosis maintained regular dental visits. Only 38.9% of a control group monitored for that time had regular dental care after 3 years.

Another common phobia that complicates medical treatment is fear of needles. Two cases exemplify how hypnosis can be useful in this area. Women's and Children's Hospital in Adelaide, Australia used hypnosis to allow treatment on a child that previously required general anesthetic to allow placement of IV lines. Tripler Army Medical Center in Honolulu, Hawaii used hypnosis to enable a lumbar puncture in a patient with dementia. It should be pointed out that in both cases, the hypnosis treatment was brief and the patient's mental ability to attend to the session was limited by their cognitive level.

The Tripler Army Medical Center also demonstrated the use of hypnosis to reduce other treatment phobias. A patient that had terminated two prior MRI procedures due to claustrophobia was able to complete a third under hypnosis.

References

Hospital Phobia: A Rapid Desensitization Technique

Waxman D. *Postgraduate Medical Journal*. 1978 May;54(631):328-30.

The less disabling phobias do not normally present a problem in that the stimulus may be avoided. This would also apply to hospital phobia until an acute medical or surgical problem might arise, when avoidance could constitute a direct threat to life. Although phobic illness is a common problem the small number of cases of hospital phobia recorded may represent the tip of the iceberg beneath which could be many phobic patients who deny their symptoms and risk their health because of their irrational fear. A case of hospital phobia in a pregnant patient with suspected disproportion was treated by a rapid desensitization technique using hypnosis. After five sessions of 30 min each, the patient was symptom free. This simple method of desensitization, if more widely known would considerably minimize the risk caused by concealment of the phobic problem.

The Use Of Hypnosis For Smooth Sedation Induction And Reduction Of Postoperative Violent Emergencies From Anesthesia In Pediatric Dental Patients

Lu DP. *ASDC Journal Of Dentistry For Children*. 1994 May-Jun;61(3):182-5.

Hypnosis combined with chemical sedation is uncommonly utilized by physicians, however, the combination of hypnosis and sedation can be an effective modality in the management of uncooperative pedodontic patients. In order to examine the efficacy of this combination technique, we selected 13 pedodontic patients for Ketamine sedation combined with hypnosis. The patients ranged from four to 11 years of age, and all had previous histories of violent emotional reactions before and after dental treatment. We found the combination technique to be extremely effective in successfully overcoming the stressful and frightening aspects of dental care for these pedodontic patients.

A 3-Year Comparison Of Dental Anxiety Treatment Outcomes: Hypnosis, Group Therapy And Individual Desensitization Vs. No Specialist Treatment

Moore R, Brødsgaard I, Abrahamsen R. *European Journal Of Oral Sciences.* 2002 Aug;110(4):287-95.

Outcomes of hypnotherapy (HT), group therapy (GT) and individual systematic desensitization (SD) on extreme dental anxiety in adults aged 19-65 yr were compared by regular attendance behaviors, changes in dental anxiety and changes in beliefs about dentists and treatment after 3 yr. Treatment groups were comparable with a static reference control group of 65 anxious patients (Dental Anxiety Scale > or = 15) who were followed for a mean of nearly 6 yr. After 3 yr, 54.5% of HT patients, 69.6% of GT patients and 65.5% of SD patients were maintaining regular dental care habits. This was better than the 46.1% of the reference group, who reported going regularly to the dentist again within the cohort follow-up period, and 38.9% of a control subgroup with observation for 3 yr. Women were better regular attenders than men at 3 yr. Specialist-treated regular attenders were significantly less anxious and had more positive beliefs than regular attenders from reference groups. There were few differences between HT, GT and SD after 3 yr. It was concluded that many patients can, on their own, successfully start and maintain regular dental treatment habits with dentists despite years of avoidance associated with phobic or extreme anxiety. However, it also appears that these patients had less success in reducing dental anxiety and improving beliefs about dentists long-term than did patients who were treated at the specialist clinic with psychological strategies.

Brief Hypnosis For Severe Needle Phobia Using Switch-Wire Imagery In A 5-Year Old

Cyna AM, Tomkins D, Maddock T, Barker D. *Paediatric Anaesthesia*. 2007 Aug;17(8): 800-4.

We present a case of severe needle phobia in a 5-year-old boy who learned to utilize a self-hypnosis technique to facilitate intravenous (i.v.) cannula placement. He was diagnosed with Bruton's disease at 5 months of age and required monthly intravenous infusions. The boy had received inhalational general anesthesia for i.v. cannulation on 58 occasions. Initially, this was because of difficult venous access but more recently because of severe distress and agitation when approached with a cannula. Oral premedication with midazolam or ketamine proved unsatisfactory and hypnotherapy was therefore considered. Following a 10-min conversational hypnotic induction, he was able to use switch--wire imagery to dissociate sensation and movement in all four limbs in turn. Two days later the boy experienced painless venepuncture without the use of topical local anesthetic cream. There was no movement in the 'switched-off' arm during i.v. cannula placement. This report adds to the increasing body of evidence that hypnosis represents a useful, additional tool that anesthetists may find valuable in everyday practice.

Use Of Hypnosis In Controlling Lumbar Puncture Distress In An Adult Needle-Phobic Dementia Patient

Simon EP, Canonico MM. *The International Journal Of Clinical And Experimental Hypnosis*. 2001 Jan;49(1):56-67.

Lumbar punctures are often vital to the medical management of patients with suspected organic pathology, yet they are commonly met with such distress that medical risk is significantly increased, and patient rapport is significantly decreased, further compromising medical treatment. Although the use of hypnosis for lumbar punctures is well established in pediatric patients, no literature exists for adult patients. Similarly, there is no extant research regarding hypnosis for dementia patients, likely due to the limiting factors of impaired attention and concentration. With these factors in mind, a method for incorporating hypnosis into a lumbar puncture procedure is described for a needle-phobic adult patient suffering from dementia.

Hypnosis Using A Communication Device To Increase Magnetic Resonance Imaging Tolerance With A Claustrophobic Patient

Simon EP. *Military Medicine*. 1999 Jan;164(1):71-2.

This is a case report of a patient who prematurely terminated two previous magnetic resonance imaging procedures because of his highly claustrophobic condition. The patient was induced into a hypnotic trance twice before his third magnetic resonance imaging examination and he was given posthypnotic suggestions for decreased anxiety and increased physiologic control. Using a communication device with headphones on the patient, he was induced into a trance as he entered the magnet. This patient was successfully able to cope with this procedure and reported great satisfaction with treatment.

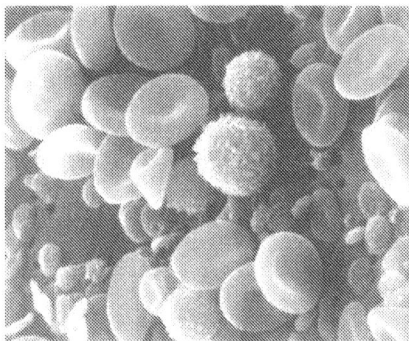

Immune Function

Summary

Almost every medical professional has seen complementary or alternative treatments making claims to "boost the immune system." As a non-specific claim this can fall outside FDA regulation and therefore mean little or nothing, depending on the treatment. Hypnosis can also make this claim but in a very specific way: hypnosis has been shown to affect the immune system measurably on a biochemical and cellular level.

Immunoglobulin A is an immune-system protein that protects mucous membranes like the inner lining of the mouth and nasal passages. Three studies have demonstrated the ability of hypnosis to significantly increase the production of this substance in the body, at Southern Methodist University, at Murdoch University in Australia, and at Minneapolis Children's

Medical Center. The Australian group also reported that the effect can help children with recurrent respiratory infections.

Histamine is another body chemical involved in local immune reactions and also in allergic reactions. At the University of Auckland in New Zealand, a hypnotic procedure reduced the sensitivity of test subjects to histamine on the skin. A group at Aarhus University Hospital in Denmark used hypnosis to induce various emotional states, and then testing the skin reactions to histamine while in those states. This latter study shows not only that emotional state affects histamine reactions, but also that hypnosis can change the body's biochemical responses.

Hypnosis can also directly affect white blood cells. Test subjects in a study at Southern Methodist University were able to decrease significantly levels of targeted types of white blood cells through hypnotic sessions.

References

Effect of Immune System Imagery On Secretory IgA

Rider MS, Achterberg J, Lawlis GF, Goven A, Toledo R, Butler JR. *Biofeedback And Self-Regulation* 1990 Dec;15(4):317-33.

This study was an investigation of the effects of physiologically-oriented mental imagery on immune functioning. College students with normal medical histories were randomly selected to one of three groups. Subjects in Group 1 participated in short educational training on the production of secretory immunoglobulin A. They were then tested on salivary IgA, skin temperature, and the Profile of Mood States (POMS) before and after listening to a 17-minute tape of imagery instructions with specially composed background "entrainment" music designed to enhance imagery. Subjects in Group 2 (placebo controls) listened to the same music but received nor formal training on the immune system. Group 3 acted as a control and subjects were tested before and after 17 minutes of no activity. Treatment groups listened to their tapes at home on a bi-daily basis for six weeks. All groups were again tested at Weeks 3 and 6. Secretory IgA was analyzed using standard radial immunodiffusion techniques. Repeated measures analyses of variance with planned orthogonal contrasts were used to evaluate the data. Significant overall increases (p less than 0.05) were found between pre- and posttests for all three trials. Groups 1 and 2 combined (treatment groups) yielded significantly greater increases in sIgA over Group 3 (control) for all three trials. Group 1 (imagery) was significantly higher than Group 2 (music) in antibody production for Trials 2 and 3. Symptomatology, recorded by subjects at Weeks 3 and 6, was significantly lower for three symptoms (rapid heartbeat, breathing difficulty, and jaw clenching), favoring both treatment groups over the control group.

Secretory Immunoglobulin A Increases During Relaxation In Children With And Without Recurrent Upper Respiratory Tract Infections

Hewson-Bower B, Drummond PD. *Journal Of Developmental And Behavioral Pediatrics*. 1996 Oct;17(5):311-6.

A diminished mucosal concentration of secretory immunoglobulin A (sIgA) in the upper respiratory tract may increase susceptibility to colds and flu. The aim of the present study was to determine whether sIgA increases during relaxation in children aged between 8 and 12 years with recurrent upper respiratory tract infections. Forty-five healthy children and 45 children with 10 or more upper respiratory tract infections in the previous year were randomly assigned to one of three experimental conditions: relaxation with suggestions to increase immune system proteins, relaxation alone, or a control condition. Samples of saliva were obtained before and after each condition. The concentration of sIgA in the saliva samples was later determined by measuring the rate of precipitation of antigen-antibody complexes to known concentration of sIgA antigen. The concentration of sIgA increased in the relaxation conditions but not in the control condition. The sIgA/albumin ratio (a more specific measure of local mucosal immunity than concentration) increased during the relaxation-suggestion condition but not during the relaxation or control conditions; however, both the concentration of sIgA and the sIgA/albumin ratio increased in proportion to subjective relaxation ratings. Neither response differed between healthy children and children with recurrent infections. The findings indicate that a disturbance in mucosal immunity in children with recurrent colds and flu does not limit increases in sIgA during relaxation. Higher preinfection levels of sIgA correlate with resistance to upper respiratory tract infection, so enhancing the sIgA concentration with relaxation techniques may help children with recurrent infection problems.

Self-Regulation Of Salivary Immunoglobulin A By Children

Olness K, Culbert T, Uden D. *Pediatrics.* 1989 Jan;83(1):66-71.

In a prospective randomized controlled study, the possibility that children could regulate their own salivary immunoglobulins was investigated using cyberphysiologic techniques. Fifty-seven children were randomly assigned to one of three groups. Group A subjects learned self-hypnosis with permission to increase immune substances in saliva as they chose; group B subjects learned self-hypnosis with specific suggestions for control of saliva immunoglobulins; group C subjects were given no instructions but received equal attention time. At the first visit, saliva samples (baseline) were collected, and each child looked at a videotape concerning the immune system and was tested with the Stanford Children's Hypnotic Susceptibility Scale. At the second visit, an initial saliva sample was collected prior to 30 minutes of self-hypnosis practice or conversation. At the conclusion of the experiment, a third saliva sample was obtained. Salivary IgA and IgG levels for all groups were stable from the first to the second sampling. Children in group B demonstrated a significant increase in IgA (P less than .01) during the experimental period. There were no significant changes in IgG. Stanford Children's Hypnotic Susceptibility Scale scores were stable across groups and did not relate to immunoglobulin changes.

Reduction In Skin Reactions To Histamine After A Hypnotic Procedure

Laidlaw TM, Booth RJ, Large RG. *Psychosomatic Medicine.* 1996 May-Jun;58(3): 242-8.

This study sought to test whether a cognitive-hypnotic intervention could be used to decrease skin reactivity to histamine and whether hypnotizability, physiological variables, attitudes, and mood would influence the size of the skin weals. Thirty eight subjects undertook three individual laboratory sessions; a pretest session to determine sensitivity to histamine, a control session, and an intervention session during which the subject experienced a cognitive-hypnotic procedure involving imagination and visualization. Compared with the control session, most subjects (32 of 38) decreased the size of their weals measured during the intervention session, and the differences between the weal sizes produced in the two sessions were highly significant (N = 38; t = 4.90; p < .0001). Mood and physiological variables but not hypnotizability scores proved to be effective in explaining the skin test variance and in predicting weal size change. Feelings of irritability and tension and higher blood pressure readings were associated with less change in weal size (i.e., a continuation of reactivity similar to that found in the control session without the cognitive-hypnotic intervention), and peacefulness and a lower blood pressure were associated with less skin reactivity during the intervention. This study has shown highly significant results in reducing skin sensitivity to histamine using a cognitive-hypnotic technique, which indicates some promise for extending this work into the clinical area.

Skin Reactions To Histamine Of Healthy Subjects After Hypnotically Induced Emotions Of Sadness, Anger, And Happiness

Zachariae R, Jørgensen MM, Egekvist H, Bjerring P. *Allergy*. 2001 Aug;56(8):734-40.

BACKGROUND: The severity of symptoms in asthma and other hypersensitivity-related disorders has been associated with changes in mood but little is known about the mechanisms possibly mediating such a relationship. The purpose of this study was to examine the influence of mood on skin reactivity to histamine by comparing the effects of hypnotically induced emotions on flare and wheal reactions to cutaneous histamine prick tests. METHODS: Fifteen highly hypnotically susceptible volunteers had their cutaneous reactivity to histamine measured before hypnosis at 1, 2, 3, 4, 5, 10, and 15 min after the histamine prick. These measurements were repeated under three hypnotically induced emotions of sadness, anger, and happiness presented in a counterbalanced order. Skin reactions were measured as change in histamine flare and wheal area in mm2 per minute. RESULTS: The increase in flare reaction in the time interval from 1 to 3 min during happiness and anger was significantly smaller than flare reactions during sadness ($P<0.05$). No effect of emotion was found for wheal reactions. Hypnotic susceptibility scores were associated with increased flare reactions at baseline ($r=0.56$; $P<0.05$) and during the condition of happiness ($r=0.56$; $P<0.05$). CONCLUSION: Our results agree with previous studies showing mood to be a predictor of cutaneous immediate-type hypersensitivity and histamine skin reactions. The results are also in concordance with earlier findings of an association between hypnotic susceptibility and increased reactivity to an allergen.

Effect Of Music-Assisted Imagery On Neutrophils And Lymphocytes

Rider MS, Achterberg J. . *Biofeedback And Self-Regulation* 1989 Sep;14(3):247-57.

The purpose of this study was to determine the effects of cell-specific mental imagery on neutrophil and lymphocyte cell counts. Subjects (N = 30) were randomly assigned to one of two experimental groups that underwent a 6-week training program focusing on images of morphology, location, and movement of either neutrophils or lymphocytes. Music was used to enhance the imagery of the subjects. Peripheral white blood cell and differential counts were determined before and after the final 20-minute imagery session. Results indicated that neutrophils decreased significantly (p less than .04) in the neutrophil-change group while lymphocytes did not. The reverse occurred in the lymphocyte-change group, with only the lymphocytes decreasing significantly (p less than .03). The authors concluded that under the conditions of the present study, cell-specific imagery was associated with decreases in peripheral blood cell counts of lymphocytes and neutrophils.

About the Authors

Esmilda Abreu-Hornbostel

Esmilda received her Master's Degree in Industrial/ Organizational Psychology from Columbia University. She is also a certified advanced instructor with the National Guild of Hypnotists.

As Director of the NeuroLink Institute, Ms. Abreu-Hornbostel has worked in association with the medical community using hypnosis, guided imagery and stress management to improve patient's mental and physical health, especially before and after surgery. The National Guild honored Ms. Abreu-Hornbostel with the "Member of the Year" award for her work in the field of hypnosis.

Ms. Abreu-Hornbostel has been featured on Eyewitness News, Newstalk Television, FX's Breakfastime, Damn Right Café, *Holistic Health Magazine* and *Good Health Naturally*.

Matthew Craver

Matthew is a writer with over ten years of experience who specializes in medical and scientific subjects. He has assisted numerous medical professionals in writing articles and book chapters, including articles on podiatry, fibromyalgia, and pain control.

www.ingramcontent.com/pod-product-compliance
Lightning Source LLC
Chambersburg PA
CBHW021824270326
41932CB00007B/328